THE TREE

"Delightful and varied, a ramble among trees he has known, from his father's overformal orchard to the last fragment of British wilderness. The real subject of this arboreal excursion is not trees at all, but the importance in art of the unpredictable, the unaccountable, the intuitive, the not discernibly useful." —*Atlantic Monthly*

"For years I have carried this book . . . with me on travels to reread, ponder, envy. In prose of classic gravity, precision, and delicacy, Fowles addresses matters of final importance. . . . His theme is the relation and the opposition between human notions of control and codification—as exemplified by his father's neatly pruned fruit trees and the nomenclature of Linnaeus on the one hand, and manifestations of the singular, the surprising, the wild, the imagination . . . on the other." —*Los Angeles Times Book Review*

"A striking display of empathy for nature and the most original argument for wilderness preservation I have encountered." —*Washington Post*

"Fowles is a master of style, evident in the ease with which he transforms the abstract into the highly tangible, without sacrificing any of the subtleties." —*Christian Science Monitor*

"Fowles's language is strong, green, discursive, related throughout to his own life and memories." —*Vogue*

"A text of unusual beauty and perception." —*Publishers Weekly*

"Gritty and entertaining." —*Sunday Telegraph* (UK)

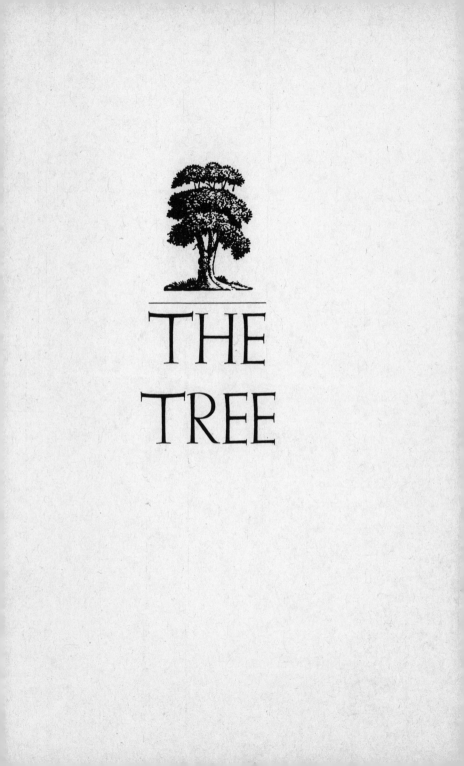

THE
TREE

ALSO BY JOHN FOWLES

THE
TREE
JOHN
FOWLES

With a New Introduction by Barry Lopez

ecco

An Imprint of HarperCollinsPublishers

HarperCollins books may be purchased for educational, business, or sales promotional use. For information, please write: Special Markets Department, HarperCollins Publishers, 10 East 53rd Street, New York, NY 10022.

An earlier version of the introduction by Barry Lopez appeared in *Sierra*.

A previous edition of this book was published in 1983 by The Ecco Press in arrangement with Little, Brown and Company.

THIS ECCO REISSUE EDITION PUBLISHED 2010

The Library of Congress has cataloged a previous edition of this book as follows:
Reproduction of the 1979 ed. by Little, Brown and Co., Boston; without the photos.
and pref. by Frank Horvat
1. Fowles, John, 1926– / —Knowledge—Natural History
2. Trees
I. TITLE
ISBN 0-88001-033-9
PR6056.085Z474 1983 823'.914 83-11555

ISBN: 978-0-06-199777-8

10 11 12 13 14   OFF/RRD   10 9 8 7 6 5 4 3 2 1

INTRODUCTION

Barry Lopez

Several times while reading this book-length essay on human perceptions of the natural world, I had to get up and walk away from it. Its thought was as stimulating as I could stand. Charged with introducing the book to other readers, I distrust this feeling. What if the thought here fits this mind too well, and order, clarity, and tone— the necessary bridges that bring these contents to other readers—are perceived to exist where they do not? The reader is then recommended a book which, without the same predilection, he or she might find dull or impenetrable.

John Fowles's graceful language, his ability to render

anecdote to illustrate abstraction—in short, his ability to tell an entertaining story—makes these concerns moot. I feel no qualm in saying Fowles has set his teeth neatly in one of the central issues of our time—our distance, real and imagined, from the natural world.

Fowles's elevated and precise prose, his almost surgical skill at teasing out the internal structure of a complex emotion or idea, and the ease with which he carries his considerable erudition may be familiar to readers of his other work—*The French Lieutenant's Woman, The Magus, Daniel Martin*, and so on. *The French Lieutenant's Woman*, among other things, is a kind of meditation on choice, on a particular facet of Darwin's theory of evolution. Indeed, Darwin and a wild wood called Ware Commons (which this book considers in passing) are protagonists of a sort in *The French Lieutenant's Woman*. Fowles has written—here, in fact—that the key to his fiction lies in his relationship with the natural world. This connection is elegant enough to almost defy conversation, but subtle, fundamental relationships are part of what Fowles is pursuing in this essay.

Who defines the natural world for us? Who decides the names of things? Who countenances and rank-orders our experience with mountains and rivers, and proceeds to define the "right" type of relationship for us to have with nature? Are we all to make some sort of effort to be what we are not—Zen contemplatives, traditional shamans, esteemed naturalists—to understand nature only in these ways? That we are too often handed others' impressions of the natural world against which to judge our own is another of Fowles's concerns here.

And how do we conceive of the natural world? As having a nonhuman purpose? As being "therapeutic" or "beautiful and enriching" and therefore useful? Do we unconsciously honor a systematized understanding of nature or any of the hundred different lexicons applied to its elements (recalling Edward Abbey's dictum: "What *is* that, madam? What it *is*, no one knows; but men call it *Artemesia tridentata*.")? Are we convinced that unless we acquire evidence—photographs, journal entries, picked and pressed flowers, sound recordings, pocketed stones— we haven't actually been intimate with nature?

Perhaps every one of us has felt such doubts and tried to thread a private, intelligent way through them, if for no other reason than that we have become innately suspicious of nature as it is popularly promoted in books and films today. In order to protect nature from exploitation we've ended up turning it into a consumer item, a repelling notion.

The validity of an individual's unarticulated experience with the natural world is what Fowles is trying to underscore here. We live at a time in which it is hard to find an unmanipulated landscape, to locate what the British writer Jay Griffiths calls "nature without an audience." The enrichment and the encouragement of a relationship with nature that might come to us through a variety of secondary sources—Bashō's *haibun* travel sketches; Pissarro's *Poplars, Eragny*; Sibelius's *The Swan of Tuonela*—are reactions we nevertheless sometimes call into question. Even with sources like these, where is the dependable authority? We're aware, too, that a way of knowing that, in Fowles's words, "leaves very little public trace," is apt not only to go unrespected but to be made the object of someone else's uninvited consolation.

Fowles was a man of considerable and deft intellect, and one of his sharpest tools of illumination here is paradox. It is just like him to both abjure, say, a film about the Arctic as "distancing" one from the actual experience of being there and to say how enriching the film is. The key to this paradox is the distinction Fowles makes between art and science. There is not the space here to elucidate, which is perhaps the coward's way out on this, but some paradoxes are forever unresolvable and therefore, like koans, provoking and valuable. The best books about nature, like this one, drive you back out there, to the inchoate, the chaotic, the unresolvable.

Fowles picks "the wood" as a focus for his thought for several reasons: what he calls its "explorability"; the social nature of trees, as well as the way they "warp time, or rather create a variety of times"; for the wood's enclosedness; and for its "uncapturability"—here Fowles is again with his genial sense of paradox. "Nowhere [but in the woods]," he writes, "are the two great contemporary modes of reproducing reality, the word and the camera, more at a loss . . . [the woods] defeat view-finder, drawing paper, canvas, they cannot be framed; and words are as futile, hopelessly too laborious and used to capture reality." But he gives us words and reveals the woods; if some intelligence one day looks back at us, it may determine it was not tool making that set us apart, or even our sense of irony, which allows us to live with paradox, but our capacity for metaphor—the way in these pages Fowles communicates the ineffable, reveals the uncapturable.

One is thankful for a gifted writer in the midst of thoughts so easily mismanaged. Like good philosophy, Fowles is in search of good argument, but because he is a storyteller we do not labor to follow him. He writes engagingly, as if he were conversing—but without hesitations or false beginnings. And his own renewal with the wild, "the refuge of the unconscious," he makes accessible to us. As with any good novel, there is structure, direction, and tension. The principal tension, a chord struck again and again, is the connection between real and imagined landscapes, between actual and metaphorical forests. We are residents of both, of course, searching for a greater resonance than we have ever been able to achieve.

Fowles opens this long, well-ordered essay by asking what separates the figurative gardeners among us from those who barely interfere with nature. He casts his father as a gardener, an inveterate pruner of limbs and puller of weeds, a man who ascribed value to nature only insofar as it yielded fruit and behaved. Fowles presents himself as the owner of an overgrown thirty-acre farm he has for the most part let go to seed. This dichotomy takes on a broader, historical dimension when Fowles introduces us to Linneaus's formal walled gardens at Uppsala. The Linnean mentality, which fussed endlessly to make nature seem categorical, serves in turn to introduce us to the differing approaches of science and "the kind of experience or knowledge we loosely define as art." Science pounces on chaos—on "unphilosophical, irrational, uncontrollable, incalculable" nature. Art perceives no threat, no great evil in unlimited chaos; the engagement with nature is personal, intimate, and without objective.

Toward the end of this section, Fowles sets down what he believes is the most dangerous of all our contemporary forms of alienation—"our growing emotional and intellectual detachment from nature." He suggests that the remedy for this lies with recognizing the debit side of the scientific revolution, understanding especially the change it has effected in our modes of perceiving and experiencing the world as individuals.

"Science is centrally, almost metaphysically, obsessed by general truths. . . . But all nature, like all humanity, is made of minor exceptions, of entities that some way, however scientifically disregardable, do not conform to the general rule. A belief in this kind of exception is as central to art as a belief in the utility of generalization is to science. . . ."

However, there is no one, true epistemology, says Fowles, not even the love of nature. "To see the woods and forests merely scientifically," he warns several pages later, "economically, topographically or aesthetically—not to understand that their greatest utility lies not in the facts derivable from them, or in their timber and fruit, or their landscape charm, or their utility as subject matter for the artist—proves the gathering speed with which we are retreating into outer space from all other life on this planet."

At the end of his essay, Fowles turns to direct experience, to what the American writer Conger Beasley calls "the unimpeachable sources." He enters a wild area of southwestern England called Wistman's Wood, where he concludes his thoughts among trees—"the best, most revealing messengers to us from all nature, the nearest to its heart." I will leave unsaid what he is able to evoke here; after pages of calm and agreeable prose, his ideas, far from being polite abstractions, are found to be rooted in the earth, abiding, preternatural and inexplicable.

This book is neither an injunction nor a call to action. It is an engaging explication of an idea before which the reader, like all readers, is alone. Toward the end of the book Fowles writes, "We still have this to learn: the inalienable otherness of each [other], human and non-human, which may seem the prison of each, but is at heart, in the deepest of the countless million metaphorical trees for which we cannot see the wood, both the justification and the redemption."

The belief that we have the power, let alone the perception, to put anything and everything into words (or images) has taught us bad habits. Fowles's entry into Wistman's Wood at the end of the book is perhaps the humblest and finest point he makes. It is also a bitter reminder that many within the broad compass of the environmental movement today are still charged by a cult of Albigensian fundamentalists in business and politics with explaining that which requires no explanation and which in fact cannot be explained. The idea that nature must be elucidated for people who have no capacity for metaphorical exegesis is, I think, the harbinger of a coming war.

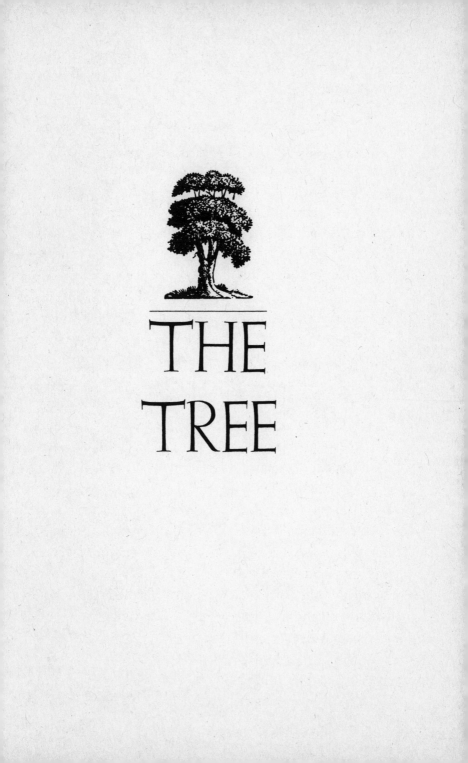

THE
TREE

The first trees I knew well were the apples and pears in the garden of my childhood home. This may sound rural and bucolic, but it was not, for the house was a semi-detached in a 1920s suburb at the mouth of the Thames, some forty miles from London. The back garden was tiny, less than a tenth of an acre, but my father had crammed one end and a side-fence with grid-iron espaliers and cordons. Even the minute lawn had five orchard apple trees, kept manageable only by constant de-branching and pruning. It was an anomaly among our neighbours' more conventional patches, even a touch absurd, as if it were trying to be a fragment of the kitchen-garden of some great country house. No one in fact thought of it as a folly, because of the fruit those trees yielded.

The names of apples and pears are rather like the names of wines—no sure guide in themselves to quality. Two labels may read the same; but the two trees that wear them may yield fruit as different as a middling and a great vineyard from the same slope. Even the same tree can vary from year to year. As with the vine, the essential things are soil, situation, annual climate; but after those chance factors, human care. My father's trees, already happy in the alluvial clay of the area, must have been among the most closely pruned, cosseted and prayed for in the whole of England, and regularly won him prizes at local shows. They were certainly the finest-flavoured of their varieties—many increasingly rare, these supermarket days, because of their commercial disadvantages, such as tender flesh or the mysterious need to be 'eaten from the tree'—that I have ever tasted. Memories of them, of their names and flavours, Charles Ross and Lady Sudeley, Peasgood's Nonsuch and King of the Pippins, haunt me still. Even the more popular kinds he grew, such as the Comice, or the Mozart and Beethoven of

English pomology, James Grieve and Cox's Orange, acquired on his cunningly stunted trees a richness and subtlety I have rarely met since. This may have been partly because he knew exactly when they should be eaten. A Comice pear may take many weeks to ripen in store, but it is at its peak for only a day. Perfection in the Grieve is almost as transient.

These trees had a far greater influence on our lives than I ever realized when I was young. I took them as my father presented them to the world, as merely his hobby; as unexceptional, or inevitable, as his constant financial worries, his disappearing every day to London, his duodenal ulcer—or on a happier side his week-end golf, his tennis, his fondness for watching county cricket. But they were already more than trees, their names and habits and characters on an emotional parity with those of family.

There was already one clear difference between my father and myself, but the child I was did not recognize it, or saw it only as a matter of taste, perhaps of age, mere choice of hobby again. The difference was in any case encouraged and in my eyes sanctified by various relatives. I had an uncle who was a keen entomologist and who took me on occasional expeditions into the country—netting, sugaring, caterpillar-hunting and all the rest of it—and taught me the delicate art of 'setting' what we caught. Then there were two cousins, much older than I was. The first was a tea planter in Kenya, a keen fly fisherman and big game shot, and indisputably to me, on his occasional home leaves and visits to us, the luckiest man in the world. The other was that indispensable member of any decent middle-class English family, a determined eccentric, who no more fitted suburbia and its values than a hedgehog fits a settee. He managed to combine a bewildering set of private interests: in vintage claret, in long-distance running (where he was of international standard), in topography, and in ants, on which he was an authority. I envied him enormously his freedom to go on walking tours, his endless photography of exotic places, his sound general field knowledge of nature, and it puzzled me that my father regarded this fascinating human being as half-mad.

What these relatives very soon aroused in me was a passion for natural history and the countryside; that is, a longing to escape from those highly unnatural trees in our back garden, and all they stood for. In this, without realizing it, I was

already trampling over my father's soul. More
and more I secretly craved everything our own
environment did not possess: space, wildness,
hills, woods . . . I think especially woodland, 'real'
trees. With one or two exceptions—the Essex
marshlands, Arctic tundra—I have always loathed
flat and treeless country. Time there seems to
dominate, it ticks remorselessly like a clock. But
trees warp time, or rather create a variety of
times: here dense and abrupt, there calm and
sinuous—never plodding, mechanical, inescap-
ably monotonous. I still feel this as soon as I enter
one of the countless secret little woods in the
Devon-Dorset border country where I now live;
it is almost like leaving land to go into water,
another medium, another dimension. When I
was younger, this sensation was acute. Slinking
into trees was always slinking into heaven.

Whether this rift would ever have developed as
it did between my father and myself if Hitler had
not been born, I cannot imagine. As it was, the
hazards of the Second World War made it inevitable. We had to leave our Essex suburb for
a remote Devonshire village, where all my secret
yearnings were to be indulged beyond my wildest
dreams. I happily forgot his little collection of

crimped and cramped fruit trees in my own new world, my America of endless natural ones in Devon. I will come to what mine meant, and mean, to me, but I must first try to convey what I now suspect his meant to him; and why they did. As I grow older I see that the outwardly profound difference in our attitudes to nature—especially in the form of the tree—had a strange identity of purpose, a kind of joint root-system, an interlacing, a paradoxical pattern.

My father was one of the generation whose lives were determined once and for all by the 1914–18 War. In most outward ways he was conventional and acutely careful not to offend the mores of the two worlds he lived in, suburbia and business London. Before the war he had trained to be a solicitor; but the death of a brother at Ypres, then of his own typically Late-Victorian father, twice married and leaving endless children to support, forced him into the tobacco trade. The family firm was nothing very grand. It specialized in Havana cigars, hand-made briar pipes, its own line of pure Virginia cigarettes (another lost flavour), and had two or three shops, including one in Piccadilly Arcade with a distinguished occasional clientèle. For various reasons—certainly not for lack of worrying on my father's part—it went into decline all through the 1930s, and the Second World War killed it for good. But every day when I was small my father, like most of his male neighbours, went off in suit and bowler to London: an hour by train there, an hour back. I very early decided that London was synonymous with physical exhaustion and nervous anxiety, and that the one thing I would never be was a commuter—a determination, I suspect, my father also held on my behalf, though for rather different reasons.

I can see now that the Great War took a doubly cruel toll on him—not only in those abominable years in the trenches, but in its social effect. He was given a taste of the life of the officer and gentleman, especially in the post-war period when he was in the occupation army in Germany. From then on he was condemned to the ethos and aspirations of a class, or way of life, that his increasingly unsuccessful business did not permit; and which our actual family background made rather absurd. My great-grandfather was clerk to an attorney in Somerset, and I think his father was a blacksmith. I like having such very ordinary ancestors, but my father, being only

a generation away from the rise out of immem-
orial West Country obscurity into well-to-do mer-
cantile London, did not. He was not a snob, he
simply hankered after a grander sort of life than
life allowed. (He did not even have the snob's out-
let of doing something about it, since he was
intensely cautious—and had to be—over money;
a trait he neither inherited from his own father
nor passed on to me.) It was far less that he believed in what we would call today upward social
mobility than that he permanently missed the
jolly expansiveness, the three-men-in-a-boatish-
ness, of a large 1890s and Edwardian household
and the style and dash of an Honorable Artillery
Company mess. None of this makes him in any
way unusual; but he had other private anomalies
beside his little sacred grove of fruit trees.

The strangest was his fascination with philosophy. That formed three-quarters of his reading, mostly in the great Germans and the
American pragmatists; the other quarter was
poetry, but again almost all of it was German
and French Romantic verse, very rarely English.
He must have known many poems of Mörike,
Droste-Hülshoff, the early Goethe, almost by
heart. Though he had one or two favourites such

as Voltaire and Daudet, the reading in French was mainly for my sake, after it had become my 'main' subject at school and at Oxford.

He virtually never read fiction; but there was a secret. It was not until I became a novelist myself that it was disclosed. My first book was well received, film rights were sold; and suddenly one day he announced to me that he had himself long ago written a novel about his war experiences, thought that it too would 'make a good film,' and asked me to read it. It was hopelessly stiff and old-fashioned, and I knew no publisher would consider it for a minute. Some of the detail of the reality of going 'over the top' in Flanders was authentic enough; and the central theme, a tale of an Englishman and his German friend in love with the same girl before the war, their coming face to face in no-man's land, death and reconciliation there, was like all baldly summarized novel themes, intrinsically neither good nor bad. But it read as if it were by someone (as it indeed was) who had read hardly a single word of all the other English fiction and poetry the Great War had produced: no Owen, no Rosenberg, no Sassoon, no Graves, no Manning . . . it was so innocent of all their sophistication, technically and emotionally, that it almost had a curiosity value, as a period piece. I asked him if he would like me to see about getting it privately printed, but he wanted his son's sort of good fortune, public acceptance and success, and I had to tell him the cruel truth.

I am sure the greatest shock for him, when I first told him I was to be published, had been the financial side; for the anomalous counterpart to this anomalous love of philosophy and Romantic verse was an obsession with yield. Just as he endlessly tended his fruit trees, so did he endlessly tend his stocks and shares in the *Financial Times* —I think probably with equal skill, though he never had very much to invest. Indeed the two things became somehow intertwined, for part of the fruit-harvesting ritual every autumn was the calculation of how much the fruit would have fetched *if* it had been sold to some local greengrocer; in fact the always considerable surplus was handed out to relations and neighbours, but I am sure this hypothetical 'dividend' was important to him. The highest praise he ever bestowed on his own produce was to say how much it had been publicly fetching the week before, as if that somehow added a cachet which superb flavour and condition could never grant. It was not the some-what scandalous—in suburban terms—content of *The Collector* that worried him nearly so much as the thought that it might be a failure; and then, when that hurdle was overcome, that I might leave the sound, if humble, economic safety-net of teaching for full-time writing. In his eyes that was like selling a blue chip for a flagrant gamble.

All I could eventually do about his own novel was to use some fragments of battlefield descrip-tion in a passage of *The Magus*. But just before he died I was sitting one afternoon by his bedside in a nursing-home; he was in pain and drugged, and seemingly asleep. Then suddenly he began to talk, a strange rattle of staccato sentences, a silence, then more, another silence. It was to do with some friend being killed beside him during an attack, and was told in terms of a dialogue between my father and some third person who had also been there. It was not in the least said for me, but came out of his near coma. There was no time, it was now again, and eternally now; infinitely more vivid in those snatches of broken sentence than in anything he had written—or indeed ever told me—of his battlefield experi-ences in his more conscious moments. They had always been taboo. Memories of Ypres and other shattered towns, chateau billets out of the line, life in occupied Cologne, yes; but never the core of it, to those who have not known it: the running, walking, plodding through wire and craters into any moment's death.

Beside his bed that day I thought of all the crossroads in our two lives where I had murdered him, or at least what he believed in, and in particular of one of his major life-decisions, never forgiven on my part, though I had long ceased to suffer from it. This went back to the end of the Second World War.

We had spent that, self-evacuated, in a cottage of the Devonshire village I fictionalized in *Daniel Martin*. Despite the external horrors and deprivations of the time, they were for me fertile and green-golden years. I learnt nature for the first time in a true countryside among true countrymen, and from then on I was irredeemably lost as a townsman. I have had to spend long years in cities since then, but never willingly, always in daily exile. I even preferred the antiquated class-system of village life, with its gentry and its 'peasants' and infinite grades between, to the uniformity of street after suburban street of same houses, same fears, same pretensions. But then, once the war was over, my father decided that we must quit the green paradise and return to the grey limbo. Neither of his overt reasons—business and the need to be near London—seemed to me honest. The family firm was virtually wound up, he had no cultural interests (unless one counts professional cricket) that were deprived by the distance of London from Devon.

I can guess now that the experience of village social distinctions, that ancient nose for the difference between status gained by money or education and status grown from ancestry, or generations of 'breeding', had upset him. But I think it was above all the breadth of choice, in terms of how and where you live, that he really disliked in the country. Some of the larger houses and gardens in our village must have corresponded to his dream; but they did not necessarily give status there. The place's most obvious gentleman (another murder at the crossroads, since he took me under his wing and taught me to fish and shoot) lived in one of its smallest cottages.

I sense that the memory of suburbia must have represented to my father (in what was also a completely new experience for him) something like the famous old fellowship of the trenches, the consoling feeling of everyone being in the same boat; all genteelly in the same reduced financial circumstances, with the same vague hopes, abiding by the same discreetly agreed codes. Things were far too transparent in Devon, too close to unfair value-systems that were in turn too close to nature; and towards nature my father showed not only no interest, but a distinct hostility (generally muted, before my own passion, into a kind of sceptical incredulity). He would claim he had seen enough open country and breathed enough open air in his three years in Flanders to last him his lifetime; and he regarded even the shortest walks, the simplest picnics away from houses and roads as incipiently dangerous, so many steps towards total anarchy. The only exception was golf, but I think that even there he regarded the rough and the surrounding woods—on the course he played—as something more than just a game hazard. The last country walk I recall attempting with him was in the Essex marshlands. He walked two or three hundred yards from where we had parked along a sea wall, then refused point-blank to go any further against a lifetime's instincts. He was old then, but quite happy still to walk two or three miles on town pavements.

He had in fact a number of the traits, both good and bad, of what used to be called the ghetto mentality: on the one hand, a keen admiration of intellectual achievement and of financial acumen (skill with yield), a love of the emotional, the Mendelssohnian, in things like poetry and classical music, of brilliant virtuoso performances (he had no time for garden plants that did not put on 'a good show'), of quintessentially city arts like the music-hall (though one of his sisters who had had the temerity to enter that last world, and once understudied Gertie Millar, was cast eternally beyond the pale); on the other, an almost total blindness to nature. This 'Jewishness' was not totally unconscious in him. Suburban neighbors who showed a stock anti-Semitism usually received very short shrift; a bludgeoning of Spinoza, perhaps, or Heine, or Einstein, and then a general lecture on what European intellectual and artistic history owed to Jewish genius. He had been a military prosecutor in Cologne, and seen most of the great Edwardian counsels in action before the war, and knew how to browbeat the shifty witness. Philosophical arguments with him could grow painfully like cross-examination, far more forensic than Socratic, and I have shunned the logical ever since.

Children are notoriously blind towards their parents, and nowhere more than in failing to see the childlike in them—the inescapable conditioning of the past. In the beginning we all try to attribute to our parents what used to be attributed to God: limitless power to intervene, indisputable wisdom. The theological concept was clearly no more than an idealization of this. Its flaw is the inevitable confusion between authority and free will—the jointly held delusion that possessing one must entail possessing the other. I am sure in retrospect that the decision to return to suburbia was well beyond my father's free will; he could not *not* do it, any more than he could prevent that terrible memory from the Great War bubbling to the surface when he lay on his deathbed. But I did at that moment guess what had truly inspired the retreat from Devon.

It was not financial caution or love of suburbia in itself, it was not anything but his trees and the sanctuary they offered . . . in no sense, in that minute garden, a physical sanctuary, but a kind of poetic one, however banal the surroundings:

a place he could control, that was different from all around it, not least in its huge annual yield of fruit. It stood in effect as the very antithesis of a battlefield, including the metaphorical one of wild nature; and of course it could not be reproduced anywhere else, since he had personally created and cherished it. We lived in Devon surrounded by farm orchards, but what he needed was the fruits of his own cultivation, the knowledge he had gained of every habit, every whim, every fruiting spur (all infertile shoots were ruthlessly extirpated) of each of his score of trees. He had himself been severely pruned by history and family circumstance, and this was his answer, his reconciliation to his fate—his platonic ideal of the strictly controlled and safe, his Garden of Eden. All my adolescent and older loathing of its social and physical environment—and my mother was on my side—can have only deepened his attachment.

Those trees were in fact his truest philosophy, and his love of actual philosophy, the world of abstract ideas, was essentially (like his love of trenchant lawyers, with secateurs in their mouths) no more than a facet of his hatred of natural disorder. Good philosophers prune the chaos of reality and train it into fixed shapes, thereby forcing it to yield valuable and delicious fruit— or at least in theory. One of my father's heroes was Bertrand Russell, for whose incisive intellect and more popular philosophical works he had the greatest admiration; yet he had the very reverse for Russell's later political attitudes. It was almost as if he had let one of his cordons grow as it liked, a blasphemous breaking of his own eleventh commandment: Thou shalt prune all trees.

I had always seen this as the great difference between us; and puzzling, genetically mysterious. What he abhorred, I adored. My own 'orchards' were, from the moments I first knew them, the forgotten and increasingly deserted copses and woods of the West of England, and later, of France. I still grow some of my father's favourite apples, such as James Grieve, and some of my own, such as the smoky King of the Pippins, but I won't use sprays and don't prune properly— with no excuse, since he taught me the rudiments of that art. Yet I see now that our very different attitudes to these things were really the same phenomenon, the same tree. His refusal to be moved by what moved me in nature was perhaps largely a product of his own conditioning; but its function (without my realizing it, of course) was very similar to what pruning does for young fruit trees—that is, to direct their growth and determine their future.

Successful artistic parents seem very rarely to give birth to equally successful artistic sons and daughters, and I suspect it may be because the urge to create, which must always be partly the need to escape everyday reality, is better fostered —despite modern educational theory—not by a sympathetic and 'creative' childhood environment, but the very opposite, by pruning and confining natural instinct. (Nine-tenths of all artistic creation derives its basic energy from the engine of repression and sublimation, and well beyond the strict Freudian definition of those terms.) That I should have differed so much from my father in this seems to me in retrospect not in the least a matter for Oedipal guilt, but a healthy natural process, just as the branches of a healthy tree do not try to occupy one another's territory. The tree in fact has biochemical and light-sensitive systems to prevent this pointless and wasteful secondary invasion of one branch's occupied space by another. The fact that the two branches grow in different directions and ways does not mean that they do not share a same mechanism of need, a same set of deeper rules.

It is immaterial that I do not cultivate trees in any sense that my father would recognize or could approve. I think I truly horrified him only once in my life, which was when, soon after coming into possession, I first took him around my present exceedingly unkempt, unmanaged and unmanageable garden. I had previously shocked him by buying a derelict farm; but its thirty acres of scrub and rough pasture were sanity (at least I let the keep and got some token yield from it) beside this new revelation of folly. He thought it madness to take on such a 'jungle', and did not believe me when I said I saw no need to take it on, only to leave it largely alone, in effect to my co-tenants, its wild birds and beasts, its plants and insects. He would never have conceded that it was my equivalent of his own beautifully disciplined apples and pears, and just as much cultivated, though not in a literal sense. He would not have understood that something I saw down there just an hour ago, at this moment I write—two tawny owlets fresh out of the nest, sitting on a sycamore branch like a pair of badly knitted Christmas stockings and ogling down at this intruder into *their* garden—means to me exactly what the Horticultural Society cups on his sideboard used to mean to him: a token of order in unjust chaos, the reward of perseverance in a right philosophy. That his chaos happens to be my order is not, I think, very important.

He sent me two cordon pear trees to plant, soon after that first visit. They must be nearly fifteen years old now; and every year, my soil being far too thin and dry for their liking, they produce a few miserable fruit, or more often none at all. I would never have them out. It touches me that they should so completely take his side; and reminds me that practically everyone else in my life—even friends who profess to be naturalists—has also taken his side; that above all the world in general continues to take his side. No fruit for those who do not prune; no fruit for those who question knowledge; no fruit for those who hide in trees untouched by man; no fruit for traitors to the human cause.

few years ago I
stood in a historic place. It was not a great battlefield, a house, a square, the site of one famous
event; but the site only of countless very small
ones—a neat little eighteenth-century garden,
formally divided by gravel walks into parterres,
with a small wooden house in one corner where
the garden's owner had once lived. There is only
one other garden to compare with it in human
history, and that is the one in the Book of
Genesis, which never existed outside words. The
one in which I stood is very real, and it lies in the
old Swedish university town of Uppsala. Its owner
was the great warehouse clerk and indexer of
nature, Carl Linnaeus, who between 1730 and
1760 docketed, or attempted to docket, most of
animate being. Perhaps nothing is more moving
at Uppsala than the actual smallness and ordered
simplicity of that garden (my father would have
loved it) and the immense consequences that
sprung from it in terms of the way we see and
think about the external world. It is something

more than another famous shrine for lovers of nature, like Selborne or Coate Farm or Walden Pond. In fact, for all its air of gentle peace, it is closer to a nuclear explosion, whose radiations and mutations inside the human brain were incalculable and continue to be so: the place where an intellectual seed landed, and is now grown to a tree that shadows the entire globe.

I am a heretic about Linnaeus, and find nothing less strange, or more poetically just, than that he should have gone mad at the end of his life. I do not dispute the value of the tool he gave to natural science—which was in itself no more than a shrewd extension of the Aristotelian system and which someone else would soon have elaborated, if he had not; but I have doubts about the lasting change it has effected in ordinary human consciousness.

It is not that I don't share some of my father's fertile attachment to the single tree, the tree in itself, and the art of cultivating it, literally or artistically. But I must confess my own love is far more of trees, more exactly of the complex internal landscapes they form when left to themselves. In the colonial organism, the green coral, of the wood or forest, experience, adventure, aesthetic pleasure, I think I could even say truth, all lie for me beyond the canopy and exterior wall of leaves, and beyond the individual.

Evolution has turned man into a sharply isolating creature, seeing the world not only anthropocentrically but singly, mirroring the way we like to think of our private selves. Almost all our art before the Impressionists—or their St John the Baptist, William Turner—betrays our love of clearly defined boundaries, unique identities, of the individual thing released from the confusion of background. This power of detaching an object from its surroundings and making us concentrate on it is an implicit criterion in all our judgements on the more realistic side of visual art; and very similar, if not identical, to what we require of optical instruments like microsopes and telescopes —which is to magnify, to focus sharper, to distinguish better, to single from the ruck. A great deal of science is devoted to this same end: to providing specific labels, explaining specific mechanisms and ecologies, in short for sorting and tidying what seems in the mass indistinguishable one from the other. Even the simplest knowledge of the names and habits of flowers or trees starts this distinguishing or individuating process, and removes us a step from total reality towards anthropocentrism; that is, it acts mentally as an equivalent of the camera view-finder. Already it destroys or curtails certain possibilities of seeing, apprehending and experiencing. And that is the bitter fruit from the tree of Uppsalan knowledge.

It also begs very considerable questions as to the realities of the boundaries we impose on what we see. In a wood the actual visual 'frontier' of any one tree is usually impossible to distinguish, at least in summer. We feel, or think we feel, nearest to a tree's 'essence' (or that of its species) when it chances to stand like us, in isolation; but evolution did not intend trees to grow singly. Far more than ourselves they are social creatures, and no more natural as isolated specimens than man is as a marooned sailor or a hermit. Their society in turn creates or supports other societies of plants, insects, birds, mammals, micro-organisms; all of which we may choose to isolate and section off, but which remain no less the ideal entity, or whole experience, of the wood—and indeed are still so seen by most of primitive mankind.

Scientists restrict the word symbiotic to those relationships between species that bring some detectable mutual benefit; but the true wood, the true place of any kind, is the sum of all its phenomena. They are all in some sense symbiotic, being together in a togetherness of beings. It is only because such a vast sum of interactions and coincidences in time and place is beyond science's calculation (a scientist might say, beyond useful function, even if calculable) that we so habitually ignore it, and treat the flight of the bird and the branch it flies from, the leaf in the wind and its shadow on the ground, as separate events, or riddles—what bird? which branch? what leaf? which shadow? These question-boundaries (where do I file that?) are ours, not of reality. We are led to them, caged by them not only culturally and intellectually, but quite physically, by the restlessness of our eyes and their limited field and acuity of vision. Long before the glass lens and the movie-camera were invented, they existed in our eyes and minds, both in our mode of perception and in our mode of analysing the perceived: endless short sequence and jump-cut, endless need to edit and range this raw material.

I spent all my younger life as a more or less orthodox amateur naturalist; as a pseudo-scientist, treating nature as some sort of intellectual puzzle, or game, in which being able to name names and explain behaviourisms—to identify and to understand machinery—constituted all the pleasures and the prizes. I became slowly aware of the inadequacy of this approach: that it insidiously cast nature as a kind of opponent, an opposite team to be outwitted and beaten; that in a number of very important ways it distracted from the total experience and the total meaning of nature —and not only of what I personally needed from nature, not only as I had long, if largely unconsciously, begun to feel it (which was neither scientifically nor sentimentally, but in a way for which I had, and still have, no word). I came to believe that this approach represented a major human alienation, affecting all of us, both personally and socially; moreover, that such alienation had much more ancient roots behind the historical accident of its present scientific, or pseudo-scientific, form.

Naming things is always implicitly categorizing and therefore collecting them, attempting to own them; and because man is a highly acquisitive creature, brainwashed by most modern societies into believing that the act of acquisition is more enjoyable than the fact of having acquired, that getting beats having got, mere names and the objects they are tied to soon become stale. There is a constant need, or compulsion, to seek new objects and names—in the context of nature, new species and experiences. Everyday ones grow mute with familiarity, so known they become un-known. And not only in non-human nature: only fools think our attitude to our fellow-men is a thing distinct from our attitude to 'lesser' life on this planet.

All this is an unhappy legacy from Victorian science, which was so characteristically obsessed with both the machine and exact taxonomy. I came only the other day on a letter in a forgotten drawer of the little museum of which I am cura-tor. It was from a well-known Victorian fern expert, concerning some twenty or so specimens he had been sent from Dorset—all reducible, to a modern botanist, to three species. But this worthy gentleman felt obliged, in a welter of Latin polysyllables, to grant each specimen some new sub-specific or varietal rank, as if they were unbaptized children and might all go to hell if they were not given individual names. It would be absurd to deny the Victorians their enormous achievements in saner scientific fields, and I am not engaging in some sort of Luddite fantasy, wishing the machine they invented had been different, or even not at all. But we are far better at seeing the immediate advantages of such gains in knowledge of the exterior world than at assessing the cost of them. The particular cost of understanding the mechanism of nature, of having so successfully itemized and pigeon-holed it, lies most of all in the ordinary person's perception of it, in his or her ability to live with and care for it —and not to see it as challenge, defiance, enemy. Selection from total reality is no less necessary in science than it is in art; but outside those domains (in both of which the final test of selection is utility, or yield, to our own species) it seriously distorts and limits any worthwhile relationship.

I caused my hosts at Uppsala, where I went to lecture on the novel, some puzzlement by demanding (the literary business once over) to see Linnaeus's garden rather than the treasures of one of the most famous libraries in Europe. The feeling that I was not behaving as a decent writer should was familiar. Again and again in recent years I have told visiting literary academics that the key to my fiction, for what it is worth, lies in my relationship with nature—I might almost have said, for reasons I will explain, in trees. Again and again I have seen, under varying degrees of politeness, this assertion treated as some sort of irrelevant quirk, eccentricity, devious evasion of what must be the real truth: literary influences and theories of fiction, all the rest of that purely intellectual midden which faculty hens and cocks so like scratching over. Of course such matters are a part of the truth; but they are no more the whole truth than that the tree we see above ground is the whole tree. Even if we do discuss nature, I soon sense that we are talking about two different things: on their side some abstract intellectual concept, and on mine an experience whose deepest value lies in the fact that it cannot be directly described by any art . . . including that of words.

One interrogator even accused me of bad faith: that if I sincerely felt so deeply on the matter, I should write more about it. But what I gain most from nature is beyond words. To try to capture it verbally immediately places me in the same boat as the namers and would-be owners of nature: that is, it exiles me from what I most need to learn. It is a little as it is in atomic physics, where the very act of observation changes what is observed; though here the catch lies in trying to describe the observation. To enter upon such a description is like trying to capture the uncapturable. Its only purpose can be to flatter the vanity of the describer—a function painfully obvious in many of the more sentimental natural history writers.

But I think the most harmful change brought about by Victorian science in our attitude to nature lies in the demand that our relation with it must be purposive, industrious, always seeking greater knowledge. This dreadfully serious and puritanical approach (nowhere better exhibited in the nineteenth century than in the countless penny magazines aimed at young people) has had two very harmful effects. One is that it turned the vast majority of contemporary Western mankind away from what had become altogether too much like a duty, or a school lesson; the second is that the far saner eighteenth-century attitude, which viewed nature as a mirror for philosophers, as an evoker of emotion, as a pleasure, a poem, was forgotten. There are intellectual reasons as well for this. Darwin made sentimental innocence, nature as mainly personal or aesthetic experience, vaguely wicked. Not only did he propose a mechanism seemingly as iron as the steam-engine, but his very method of discovery, and its success in solving a great conundrum, offered an equally iron or one-sided model for the amateur naturalist himself, and made the older and more humanist approach seem childish. A 'good' amateur naturalist today merely means one whose work is valued by the professional scientists in his field.

An additional element of alienation has come with the cinema and television, which are selective in another way. They present natural reality not only through other eyes, but a version of it in which the novelty or rarity of the subject plays a preponderant part in choice and treatment. Of course the nature film or programme has an entertainment value; of course there are some social goods in the now ubiquitous availability of copies of other people's images and opinions of actual things and events; but as with the Linnaean system, there is a cost. Being taken by camera into the deepest African jungle, across the Arctic wastes, thirty fathoms deep in the sea, may seem a 'miracle of modern technology'; but it will no more bring the viewer nearer the reality of nature, or a proper human relationship with the actual nature around him, than merely reading novels is likely to teach the writing of them. The most one can say is that it may help; a much more common result is to be persuaded of the futility of even trying.

Increasingly we live (and not only in terms of nature and novels) by the old tag, *Aut Caesar, aut nullus*. If I can't be Caesar, I'll be no one. If I can't have the knowledge of a scientist, I'll know nothing. If I can't have superb close-ups and rare creatures in the nature around me, to hell with it. Perhaps any representation of nature is better, to those remote from it in their daily lives, than none. Yet a great deal of such representation seems to me to descend straight from the concept of the menagerie, another sadly alienating selection, or reduction, from reality. Poking umbrellas through iron bars did not cease with the transition from the zoo to the screen.

Much of seventeenth- and eighteenth-century science and erudition is obsolete nonsense in modern scientific terms: in its personal interpolations, its diffuse reasoning, its misinterpreted evidence, its frequent blend of the humanities with science proper—its quotations from Horace and Virgil in the middle of a treatise on forestry. But one general, if unconscious, assumption lying behind almost all pre-Victorian science—that it is being presented by an entire human being, with all his complexities, to an audience of other entire human beings—has been much too soon dismissed as a mere historical phenomenon, at best exhibiting an engaging amateurishness, at worst sheer stupidity, from neither of which we have anything to learn. It is not of course the fault of modern scientists that most of their formal discourse is now of so abstruse a nature that only their fellow specialists can hope to understand it; that the discourse itself is increasingly mechanical, with words reduced to cogs and treated as poor substitutes for some more purely scientific formulation; nor is it directly their fault that their vision of empirical knowledge, the all-important value they put upon proven or demonstrable fact, has seeped down to dominate the popular view of nature—and our education about it. Our fallacy lies in supposing that the limiting nature of scientific method corresponds to the nature of ordinary experience.

Ordinary experience, from waking second to second, is in fact highly synthetic (in the sense of combinative or constructive), and made of a complexity of strands, past memories and present perceptions, times and places, private and public history, hopelessly beyond science's powers to analyse. It is quintessentially 'wild', in the sense my father disliked so much: unphilosophical, ir-rational, uncontrollable, incalculable. In fact it corresponds very closely—despite our endless ef-forts to 'garden', to invent disciplining social and intellectual systems—with wild nature. Almost all the richness of our personal existence derives from this synthetic and eternally present 'con-fused' consciousness of both internal and external reality, and not least because we know it is beyond the analytical, or destructive, capacity of science.

Half by its principles, half by its inventions, science now largely dictates and forms our com-mon, or public, perception of and attitudes to external reality. One can say of an attitude that it is generally held by society; but society itself is an abstraction, a Linnaeus-like label we apply to a group of individuals seen in a certain context and for a certain purpose; and before the attitude can be generally held, it must pass through the filter of the individual consciousness, where this irreducible 'wild' component lies—the one that may agree with science and society, but can never be wholly plumbed, predicted or commanded by them.

One of the oldest and most diffused bodies of myth and folklore has accreted round the idea of the man in the trees. In all his manifestations, as dryad, as stag-headed Herne, as outlaw, he possesses the characteristic of elusiveness, a power of 'melting' into the trees, and I am certain the attraction of the myth is so profound and universal because it is constantly 'played' inside every individual consciousness.

This notion of the green man—or green woman, as W. H. Hudson made her—seen as emblem of the close connection between the actuality of present consciousness (not least in its habitual flight into a mental greenwood) and what science has censored in man's attitude to nature—that is, the 'wild' side, the inner feeling as opposed to the outer, fact-bound, conforming face imposed by fashion—helped me question my old pseudo-scientist self. But it also misled me for a time. In the 1950s I grew interested in the Zen theories of 'seeing' and of aesthetics: of learning to look beyond names at things-in-themselves. I stopped bothering to identify species new to me, I concentrated more and more on the familiar, daily nature around me, where I then lived. But living without names is impossible, if not downright idiocy, in a writer; and living without explanation or speculation as to causality, little better—for Western man, at least. I discovered, too, that there was less conflict than I had imagined between nature as external assembly of names and facts and nature as internal feeling; that the two modes of seeing or knowing could in fact marry and take place almost simultaneously, and enrich each other.

Achieving a relationship with nature is both a science and an art, beyond mere knowledge or mere feeling alone; and I now think beyond oriental mysticism, transcendentalism, 'meditation techniques' and the rest—or at least as we in the West have converted them to our use, which seems increasingly in a narcissistic way: to make ourselves feel more positive, more meaningful, more dynamic. I do not believe nature is to be reached that way either, by turning it into a therapy, a free clinic for admirers of their own sensitivity. The subtlest of our alienations from it, the most difficult to comprehend, is our need to use it in some way, to derive some personal yield. We shall never fully understand nature (or ourselves), and certainly never respect it, until we dissociate the wild from the notion of usability—however innocent and harmless the use. For it is the general uselessness of so much of nature that lies at the root of our ancient hostility and indifference to it.

There is a kind of coldness, I would rather say a stillness, an empty space, at the heart of our forced co-existence with all the other species of the planet. Richard Jefferies coined a word for it: the ultra-humanity of all that is not man . . . not with us or against us, but outside and beyond us, truly alien. It may sound paradoxical, but we shall not cease to be alienated—by our knowledge, by our greed, by our vanity—from nature until we grant it its unconscious alienation from us.

I am not one of those supreme optimists who think all the world's ills, and especially this growing divide between man and nature, can be cured by a return to a quasi-agricultural, ecologically 'caring' society. It is not that I doubt it might theoretically be so cured; but the possibility of the return defeats my powers of imagination. The majority of Western man is now urban, and the whole world will soon follow suit. A very significant tilt of balance in human history is expected by the end of the coming decade: over half of all mankind will by then have moved inside towns and cities. Any hope of reversing that trend, short of some universal catastrophe, is as tiny and precarious as the Monarch butterflies I watched, an autumn or two ago, migrating between the Fifth Avenue skyscrapers in central Manhattan. All chance of a close acquaintance with nature, be it through intellect and education, be it in the simplest way of all, by having it near at hand, recedes from the many who already effectively live in a support system in outer space, a creation of science, and without means to escape it, culturally or economically.

But the problem is not, or only minimally, that nature itself is in imminent danger or that we shall lose touch with it simply because we have less access to it. A number of species, environments, unusual ecologies are in danger, there are major pollution problems; but even in our most densely populated countries the ordinary wild remains far from the brink of extinction. We may not exaggerate the future threats and dangers, but we do exaggerate the present and actual state of this global nation—under-estimate the degree to which it is still surviving and accessible to those who want to experience it. It is far less nature itself that is yet in true danger than our attitude to it. Already we behave as if we live in a world that holds only a remnant of what there actually is; in a world that may come, but remains a black hypothesis, not a present reality.

I believe the major cause of this more mental than physical rift lies less in the folly or onesidedness of our societies and educational systems, or in the historical evolution of man into a predominantly urban and industrial creature, a thinking termite, than in the way we have, during these last hundred and fifty years, devalued the kind of experience or knowledge we loosely define as art; and especially in the way we have failed to grasp its deepest difference from science. No art is truly teachable in its essence. All the knowledge in the world of its techniques can provide in itself no more than imitations or replicas of previous art. What is irreplaceable in any object of art is never, in the final analysis, its technique or craft, but the personality of the artist, the expression of his or her unique and individual feeling. All major advances in technique have come about to serve this need. Techniques in themselves are always reducible to sciences, that is, to learnability. Once

Joyce has written, Picasso painted, Webern composed, it requires only a minimal gift, besides patience and practice, to copy their techniques exactly; yet we all know why this kind of technique-copy, even when it is so painstakingly done—for instance, in painting—that it deceives museum and auction-house experts, is counted worthless beside the work of the original artist. It is not *of* him or her, it is not art, but imitation.

As it is with the true 'making' arts, so it is with the other aspects of human life of which we say that full knowledge or experience also requires an art—some inwardly creative or purely personal factor beyond the power of external teaching to instil or science to predict. Attempts to impart recipes or set formulae as to practice and enjoyment are always two-edged, since the question is not so much whether they may or may not enrich the normal experience of that abstract thing, the normal man or woman, but the certainty that they must in some way damage that other essential component of the process, the contribution of the artist in this sense—the individual experiencer, the 'green man' hidden in the leaves of his or her unique and once-only being.

Telling people why, how and when they ought to feel this or that—whether it be with regard to the enjoyment of nature, of food, of sex, or anything else—may, undoubtedly sometimes does, have a useful function in dispelling various kinds of socially harmful ignorance. But what this instruction cannot give is the deepest benefit of any art, be it of making, or of knowing, or of experiencing: which is self-expression and self-discovery. The last thing a sex-manual can be is an *ars amoris*—a science of coupling, perhaps, but never an art of love. Exactly the same is true of so many nature-manuals. They may teach you how and what to look for, what to question in external nature; but never in your own nature.

In science greater knowledge is always and indisputably good; it is by no means so throughout all human existence. We know it from art proper, where achievement and great factual knowledge, or taste, or intelligence, are in no way essential companions; if they were, our best artists would also be our most learned academics. We can know it by reducing the matter to the absurd, and imagining that God, or some Protean visitor from outer space, were at one fell swoop to grant us all knowledge. Such omniscience would be worse than the worst natural catastrophe, for our species as a whole; would extinguish its soul, lose it all pleasure and reason for living.

This is not the only area in which, like the rogue computer beloved of science fiction fans, some socially or culturally consecrated proposition—which may be true or good in its social or cultural context—extends itself to the individual; but it is one of the most devitalizing. Most mature artists know that great general knowledge is more a hindrance than a help. It is only innately mechanical, salami-factory novelists who set such great store by research; in nine cases out of ten what natural knowledge and imagination cannot supply is in any case precisely what needs to be left out. The green man in all of us is well aware of this. In practice we spend far more time rejecting knowledge than trying to gain it, and wisely. But it is in the nature of all society, let alone one deeply imbued with a scientific and technological ethos, to bombard us with ever more knowledge —and to make any questioning or rejection of it unpatriotic and immoral.

Art and nature are siblings, branches of the one tree; and nowhere more than in the continuing inexplicability of many of their processes, and above all those of creation and of effect on their respective audiences. Our approach to art, as to nature, has become increasingly scientized (and dreadfully serious) during this last century. It sometimes seems now as if it is principally there not for itself but to provide material for labeling, classifying, analysing—specimens for 'setting', as I used to set moths and butterflies. This is of course especially true of—and pernicious in—our schools and universities. I think the first sign that I might one day become a novelist (though I did not then realize it) was the passionate detestation I developed at my own school for all those editions of examination books that began with a long introduction: an anatomy lesson that always reduced the original text to a corpse by the time one got to it, a lifeless demonstration of a pre-established proposition. It took me years to realize that even geniuses, the Shakespeares, the Racines, the Austens, have human faults.

Obscurity, the opportunity a work of art gives for professional explainers to show their skills, has become almost an aesthetic virtue; at another extreme the notion of art as vocation (that is, something to which one is genetically suited) is dismissed as non-scientific and inegalitarian. It is not a gift beyond personal choice, but one that can be acquired, like knowledge of science, by rote, recipe and hard work. Elsewhere we become so patterned and persuaded by the tone of the more serious reviewing of art in our magazines and newspapers that we no longer notice their overwhelmingly scientific tone, or the paradox of this knowing-naming technique being applied to a non-scientific object—one whose production the artist himself cannot fully explain, and one whose effect the vast majority of the non-reviewing audience do not attempt to explain.

The professional critic or academic would no doubt say this is mere ignorance, that both artists and audiences have to be taught to understand themselves and the object that links them, to make the relationship articulate and fully conscious; defoliate the wicked green man, hunt him out of his trees. Of course there is a place for the scientific, or quasi-scientific, analysis of art, as there is (and far greater) for that of nature. But the danger, in both art and nature, is that all emphasis is placed on the created, not the creation.

All artefacts, all bits of scientific knowledge, share one thing in common: that is, they come to us from the past, they are relics of something already observed, deduced, formulated, created, and as such qualify to go through the Linnaean and every other scientific mill. Yet we cannot say that the 'green' or creating process does not happen or has no importance just because it is largely private and beyond lucid description and rational analysis. We might as well argue that the young wheat-plant is irrelevant because it can yield nothing to the miller and his stones. We know that in any sane reality the green blade is as much the ripe grain as the child is father to the man. Nor of course does the simile apply to art alone, since we are all in a way creating our future out of our present, our 'published' outward behaviour out of our inner green being. One main reason we may seldom feel this happening is that society does not want us to. Such random personal creativity is offensive to all machines.

I began this wander through the trees—we shall come to them literally, by the end—in search of that much looser use of the word 'art' to describe a way of knowing and experiencing and enjoying outside the major modes of science and art proper . . . a way not concerned with scientific discovery and artefacts, a way that is internally rather than externally creative, that leaves very little public trace; and yet which for those very reasons is almost wholly concentrated in its own creative process. It is really only the qualified scientist or artist who can escape from the interiority and constant nowness, the green chaos of this experience, by making some aspect of it exterior and so fixing it in past time, or known knowledge. Thereby they create new, essentially parasitical orders and categories of phenomena that in turn require both a science and an art of experiencing.

But nature is unlike art in terms of its product —what we in general know it by. The difference is that it is not only created, an external object with a history, and so belonging to a past; but also creating in the present, as we experience it. As we watch, it is so to speak rewriting, reformulating, repainting, rephotographing itself. It refuses to stay fixed and fossilized in the past, as both the scientist and the artist feel it somehow ought to; and both will generally try to impose this fossilization on it.

Verbal tenses can be very misleading here: we stick adamantly in speech to the strict protocol of actual time. Of and in the present we speak in the present, of the past in the past. But our psychological tenses can be very different. Perhaps because I am a writer (and nothing is more fictitious than the past in which the first, intensely alive and present, draft of a novel goes down on the page), I long ago noticed this in my naturalist self: that is, a disproportionately backward element in any present experience of nature, a retreat or running-back to past knowledge and experience, whether it was the definite past of personal memory or the indefinite, the imperfect, of stored 'ological' knowledge and proper scientific behaviour. This seemed to me often to cast a mysterious veil of deadness, of having already happened, over the actual and present event or phenomenon.

I had a vivid example of it only a few years ago in France, long after I thought I had grown wise to this self-imposed brain-washing. I came on my first Soldier Orchid, a species I had long wanted to encounter, but hitherto never seen outside a book. I fell on my knees before it in a way that all botanists will know. I identified, to be quite certain, with Professors Clapham, Tutin and Warburg in hand (the standard British *Flora*), I measured, I photographed, I worked out where I was on the map, for future reference. I was excited, very happy, one always remembers one's 'firsts' of the rarer species. Yet five minutes after my wife had finally (other women are not the only form of adultery) torn me away, I suffered a strange feeling. I realized I had not actually *seen* the three plants in the little colony we had found. Despite all the identifying, measuring, photographing, I had managed to set the experience in a kind of present past, a having-looked, even as I was temporally and physically still looking. If I had the courage, and my wife the patience, I would have asked her to turn and drive back, because I knew I had just fallen, in the stupidest possible way, into an ancient trap. It is not necessarily too little knowledge that causes ignorance; possessing too much, or wanting to gain too much, can produce the same result.

There is something in the nature of nature, in its presentness, its seeming transience, its creative ferment and hidden potential, that corresponds very closely with the wild, or green man, in our psyches; and it is a something that disappears as soon as it is relegated to an automatic pastness, a status of merely classifiable *thing*, image taken *then*. 'Thing' and 'then' attract each other. If it is thing, it was then; if it was then, it is thing. We lack trust in the present, this moment, this actual seeing, because our culture tells us to trust only the reported back, the publicly framed, the edited, the thing set in the clearly artistic or the clearly scientific angle of perspective. One of the deepest lessons we have to learn is that nature, of its nature, resists this. It waits to be seen otherwise, in its individual presentness and from our individual presentness.

I come now near the heart of what seems to me to be the single greatest danger in the rich legacy left us by Linnaeus and the other founding fathers of all our sciences and scientific mores and methods—or more fairly, left us by our leaping evolutionary ingenuity in the invention of tools. All tools, from the simplest word to the most advanced space probe, are disturbers and rearrangers of primordial nature and reality—are, in the dictionary definition, 'mechanical implements for working upon something.' What they have done, and I suspect in direct proportion to our ever-increasing dependence on them, is to addict us to purpose: both to looking for purpose in everything external to us and to looking internally for purpose in everything we do—to seek explanation of the outside world by purpose, to justify our seeking by purpose. This addiction to finding a reason, a function, a quantifiable yield, has now infiltrated all aspects of our lives—and become effectively synonymous with pleasure. The modern version of hell is purposelessness.

Nature suffers particularly in this, and our indifference and hostility to it is closely connected with the fact that its only purpose appears to be being and surviving. We may think that this comprehends all animate existence, including our own; and so it must, ultimately; but we have long ceased to be content with so abstract a motive. A scientist would rightly say that all form and behaviour in nature is highly purposive, or strictly designed for the end of survival—specific or genetic, according to theory. But most of this functional purpose is hidden to the non-scientist, indecipherable; and the immense variety of nature appears to hide nothing, nothing but a green chaos at the core—which we brilliantly purposive apes can use and exploit as we please, with a free conscience.

A green chaos. Or a wood.

In some mysterious way woods have never seemed to me to be static things. In physical terms, I move through them; yet in metaphysical ones, they seem to move through me, just as, if I watch a film, I stay physically in one place and it is the images in the projector gate that shift; as do the words on the page and the scenes they evoke, when read. This inner or mental reversal of the actual movement, common to all traveling, comes very close to what I like most in all narrative art from the novel to the cinema: that is, the motion from a seen present to a hidden future. The reason that woods provide this experience so naturally and intensely lies, of course, in the purely physical character of any large congregation of trees; in the degree to which they hide what exists, at any given point, beyond the immediately visible surroundings. In this they are like series of rooms and galleries, house-like, doored and screened, continuous yet separate; or paged and chaptered, like a fiction. Just as with fiction, there are in this sense good and bad tree congregations—some that tempt the visitor to turn the page, to explore further, others that do not. But even the most 'unreadable' woods and forests are in fact subtler than any conceivable fiction, which can never represent the actual multiplicity of choice of paths in a wood, but only one particular path through it. Yet that multiplicity of choice, though it cannot be conveyed in the frozen medium of the printed text, is very characteristic of the actual writing; of the constant dilemma—pain or pleasure, according to circumstances—its actual practice represents, from the formation of the basic sentence to the larger matters of narrative line, character development, ending. Behind every path and every form of expression one does finally choose, lie the ghosts of all those that one did not.

I do not plan my fiction any more than I normally plan woodland walks; I follow the path that seems most promising at any given point, not some itinerary decided before entry. I am quite sure this is not some kind of rationalization, or irrationalization, after the fact; that having discovered I write fiction in a disgracefully haphazard sort of way, I now hit on the passage through an unknown wood as an analogy. It is the peculiar nature of my adolescent explorings of the Devon countryside (peculiar because I had not been brought up in a rural atmosphere, could not take the countryside for granted, indeed it came to me with something of the unreality, the not-quite-thereness of a fiction) that made me what I am—and in many other ways besides writing.

I see now that what I liked best about the green density, the unpeopled secrecy of the Devon countryside that the chances of history gave me was its explorability. At the time I thought I was learning to shoot and fish (also to trespass and poach, I am afraid), to botanize and bird-watch; but I was really addicting myself, and beyond curability, to the pleasures of discovery, and in particular of isolated discovery and experience. The lonelier the place, the better it pleased me: its silence, its aura, its peculiar conformation, its enclosedness. I had a dream of some endless combe, I suppose almost an animal dream, an otter-dream, of endless hanging beech-woods and hazel-coppices and leated meadows, houseless and manless. It was not quite without substance in those days, such 'lost' valleys still existed and in some of them the rest of the world did not. But of course they were finite, and at some point ended at a lane, a cottage or farmhouse, 'civilization'; and discovery died.

The cost of all this is that I have never gained any taste for what lies beyond the experience of solitary discovery—in terms of true geographical exploration, for the proper exploitation of the discovery. I have dabbled in many branches of natural (and human) history, and have a sound knowledge of none; and the same goes for countless other things besides. I like a kind of wandering wood acquaintance, and no more; a dilettante's, not a virtuoso's; always the green chaos rather than the printed map. I have method in nothing, and powers of concentration, of patience in acquiring true specialized knowledge, that would disgrace a child. I can concentrate when I write, but purely because it is a sublimated form of discovery, isolated exploration, my endless combe in leaves of paper. I place all this entirely upon the original adolescent experience, for I do not think I was born so, with a painfully low threshold of boredom before learning or knowledge that is not clearly assimilable to the experience of solitary discovery.

Perhaps because I was brought up without any orthodox faith, and remain without it, there was also, I suspect, some religious element in my feeling towards woods. Their mysterious atmospheres, their silences, the parallels—especially in beechwoods—with columned naves that Baudelaire seized on in his famous line about a temple of living pillars, all these must recall the man-made holy place. We know that the very first holy places in Neolithic times, long before Stonehenge (which is only a petrified copse), were artificial wooden groves made of felled, transported and re-erected tree trunks; and that their roofs must have seemed to their makers less roofs than artificial leaf-canopies. Even the smallest woods have their secrets and secret places, their unmarked precincts, and I am certain all sacred buildings, from the greatest cathedral to the smallest chapel, and in all religions, derive from the natural aura of certain woodland or forest settings. In them we stand among older, larger and infinitely other beings, remoter from us than the most bizarre other non-human forms of life: blind, immobile, speechless (or speaking only Baudelaire's *confuses paroles*), waiting . . . altogether very like the only form a universal god could conceivably take. The Neolithic peoples, the slaves, as we are of an industrial economy, of their own great new cultural 'invention' of farming, were the first great deforesters of our landscapes, and perhaps it was guilt that made them return to the trees to find a model for their religious buildings—in which they were followed by the Bronze Age, the Greeks and Romans with their columns and porticoes, the Celtic Iron Age with its Druids and sacred oak-groves.

There is certainly something erotic in them, as there is in all places that isolate and hide; but woods are in any case highly sensuous things. They may not carry more species than some other environments, but they are far richer and more dramatic in sensory impressions. Nowhere are the two great contemporary modes of reproducing reality, the word and the camera, more at a loss; less able to capture the sound (or soundlessness) and the scents, the temperatures and moods, the all-roundness, the different levels of being in the vertical ascent from ground to tree-top, in the range of different forms of life and the subtlety of their inter-relationships. In a way woods are like the sea, sensorially far too various and immense for anything but surfaces or glimpses to be captured. They defeat view-finder, drawing-paper, canvas, they cannot be framed; and words are as futile, hopelessly too laborious and used to capture the reality.

Lt is not for nothing that the ancestors of the modern novel that began to appear in the early Middle Ages so frequently had the forest for setting and the quest for central theme. Every novel since literary time began, since the epic of Gilgamesh, is a form of quest, or adventure. Only two other environments can match the forest as setting for it; and even then, not very favourably. The horizontality of the sea hides too little. The only screen in outer space is space itself. They are also much remoter from our human scale, their vistas far less immediately and incessantly curtailed. Never mind that the actual forest is often a monotonous thing, the metaphorical forest is constant suspense, stage awaiting actors: heroes, maidens, dragons, mysterious castles at every step.

It may be useless as a literal setting in an age

that has lost all belief in maidens, dragons and magical castles, but I think we have only superficially abandoned the basic recipe (danger, eroticism, search) first discovered by those early medieval writers. We have simply transferred the tree setting to the now more familiar brick-and-concrete forest of town and city. Certain juxtapositions of tree and building, especially in city hearts, and perhaps most strikingly of all in New York, have always rather touched me: the sight of those literal and symbolic leaf-walls standing side by side, half-hiding, half-revealing, can be strangely poetic, and not just in architectural terms. Older and less planned quarters of cities and towns are profoundly woodlike, and especially in this matter of the mode of their passage through us, the way they unreel, disorientate, open, close, surprise, please. The stupidest mistake of all the many stupid mistakes of twentieth-century architecture has been to forget this ancient model in the more grandiose town-planning. Geometric, linear cities make geometric, linear people; wood cities make human beings.

This assertion would have seemed very near heresy to the medieval mind, and politically dangerous to those of the Renaissance and seventeenth-century Europe. The attraction of the forest setting to the early pioneers of fiction was in no way an attraction to the forest itself. It was clearly evil; but being evil, gave convenient excuse for the legitimate portrayal of all its real or supposed dangers to the traveller. The church might complain about the eagerness with which the educated public throughout Europe took to these tree-tales of adultery, magic, mystery, monsters, eternal danger and eternal temptation. But it could hardly deny the general truth of a proposition it was itself increasingly determined to maintain: the inherent wickedness of godless nature, in outer reality as in man himself. Raymond Chandler and the other creators of our own century's private eyes have used exactly the same technique, substituting evil city for evil trees and then giving themselves a comprehensive licence, behind the pretext of an incorruptible hero, to describe all the vices, horrors and seductions from the straight path whose gauntlet he has to run in order to earn the adjective. Sir Galahad and Philip Marlowe are blood brothers.

During very nearly all of the last thousand years true human virtue (and virtuous beauty) has lain for European mankind in nature tamed, on its knees inside the *hortus conclusus*, or emblematic walled garden of civilization. So powerful was this concept that naturalistic artistic representation of wild landscape is entirely absent before the seventeenth century, and so rare then that one might almost say, before the advent of the Romantic Movement; while public concern for nature, with positive steps to protect it, did not come until well into the nineteenth century and even then only very intermittently. Our own, the last of the millennium, is in fact the first to show some sort of general and international concern; and I do not think we should be too self-congratulatory about that. The future may well judge that we had both the scientific awareness and the political organization, the potential, to do much more than we have done.

Nor is it simply that in the medieval beginnings of our suspicious attitude to nature so many artists employed the literal imagery of the garden of Eden, of Paradise, of Virgin and docile unicorn in a bower. Even when wilderness and chaos —the two were virtually synonymous—had to be shown in such things as the backgrounds of hermit and hell pictures, they were as formally arranged, as parklike, as the closed garden itself;

exactly as if the physical limits of the painting were metaphorical garden walls, and nothing inside could be presented as it really existed, behaved and grew. Of course this high formality now seems to us one of the great charms of medieval art, and one cannot blame the earlier medieval artists for failing to put down what they in any case lacked the techniques to represent, even if they had possessed the wish and the clear vision.

But those techniques came, and it seems to me that nothing is more revealing than the inability of such artists as Pisanello and Dürer to compass the reality of the wild—for all their determination in other things, such as human portraiture, to look nature entire in the face. Clearly two such sharp observers and superb draughtsmen could technically have conveyed it; yet some deep mental blindness, or complex, prevented them. Dürer's tuft of violets or his hare, Pisanello's lizards, stags, his hoopoe and his cheetah (surely the most beautiful single drawing ever done of that animal) may seem to us as 'natural', as realistic as a modern photograph. But in terms of art history they must also seem surreptitious, bearing a faint stigma of the pornographic, of a secret wickedness the more public artist had to deny; for as soon as such individual elements become no more than components in a wider scene, they must be gardened, artificially posed and arranged, turned into mere emblems.

We all have our favourite pictures, or ikons, and one of mine has long been a painting by Pisanello in the National Gallery in London, *The Vision of St. Eustace*. The saint-to-be sits on his horse in a forested wilderness—he is out hunting —arrested before his vision of a stag bearing Christ crucified between its antlers. Other animals, birds and flowers crowd the background of the small picture. The artifice of the ensemble, above all when compared with Pisanello's own survived work-sketches of individual beast and bird in it, is almost total. The sketches and drawings are entirely and dazzlingly naturalistic; yet in the painting their subjects become as heraldic and symbolic, as unreally juxtaposed, as beasts in a tapestry. I know no picture that demonstrates more convincingly, and touchingly, this strange cultural blindness; and it is fitting that Pisanello should have chosen the patron saint of dogs (and formerly of hunting, before St Hubert usurped that role) as the central figure, and distorter of the non-human life around him. What is truly being hounded, harried and crucified in this ambiguous little masterpiece is not Christ, but nature itself.

Even the great seventeenth-century landscapists, such as Ruysdael, do not really get close to natural reality, if one compares their portrayal of it with that of contemporary towns and other human artefacts; it was still mere background to be composed and gardened in accordance with their own notion of the picturesque—far less treescapes, in a painter like Hobbema, than townscapes composed with trees instead of houses. Nature by then was not so much to be feared and anathematized as slighted and mistrusted: to be improved, made tasteful. In many ways painters did not begin to see nature whole until the camera saw it for them; and already, in this context, had begun to supersede them.

Art has no special obligation to be realistic and naturalistic, indeed any obligation at all except to say what the artist wants or chooses to say. Yet this long-lasting inability to convey the whole as truthfully as the isolated part—this failure to match the human eye (or the camera) in the ensemble, despite having equalled it in detail at least four centuries (Pisanello died in 1455) before the camera's invention—is symptomatic of a long and damaging doubt in man.

There are very understandable practical reasons why well into the sixteenth century European man (at home as well as on his voyages of exploration) should regard untamed nature very much as he regarded the sea—as a vast and essentially hostile desert, a kind of necessary evil. Commerce, personal profit, government, social stability and many other things required that the then largely arboreal wasteland between towns and cities should be crossed; but there was no pleasure in it, beyond safe arrival the other end . . . except perhaps that of hunting; but even that was the sport of a few, and done armed, in safe parties.

As in so much else, the Robin Hood myth, or that part of it that suggests life under the greenwood tree can be pleasant, runs profoundly counter to the general feeling and spirit of the Middle Ages; and even in the Robin Hood corpus, the happy greenwood side is much more an element of the Elizabethan and later ballads and accounts than the earlier ones. It is probably no coincidence that the end of the first great wave of common-land enclosure and the rise of the Puritan ethos both took place in Elizabethan times. The first hints of a rebellious and irreligious swing from nature-fearing to nature-liking took place then. The pastoral settings and themes of some of Shakespeare's plays—the depiction of not totally unrewarding exiles from the safe garden of civilization in A *Midsummer Night's Dream, As You Like It, The Tempest* and the rest—are not examples of the foresight of genius, but skilful pandering to a growing vogue. Yet little of this is reflected in actual seventeenth-century ways of life—and least of all in its gardens, which remained in general quite as formal as medieval ones. Nature still remained a potential dissolver of decency, a notion that the endless chain of new discoveries about the ways of more primitive man—the nearer nature, the nearer Caliban—did nothing to dispel. It remained essentially an immense green cloak for Satan: for the commission of crime and sin, for doubters of religious and public order; above all for impious doubters of man himself, as God's chosen steward and bailiff over the rest of creation.

We may think, now that the steward has so

comprehensively reversed the old ratio of nature
to civilization, that such superstitious hatreds and
fears of the wild are dead—and especially their
indispensable corollary, the idea of all virtue and
beauty lying inside the confines of the *hortus
conclusus*. But I see little sign of it, and certainly
not in the way ordinary householders in Europe
and America still run their own gardens—or in
the considerable industry that supplies their
needs in terms of pesticides and herbicides. The
one place—and ominously close to us, both
physically and psychologically—in which wild
nature remains unwelcome and detested is the
private garden; and this is despite its growing popularity in terms of books and television screen,
and all the endeavours of the conservationists.

I remember a strange event, in that suburban
road in Essex where I was born. One of the
elderly residents went slightly mad on the death
of his wife; he drew his curtains and turned his
back on the outside world. There was at first
considerable sympathy for the poor man, until
it was realized that the outside world included his
own garden. No grass was cut, no beds weeded,
no trees pruned; the place ran riot with dandelion,
ragwort, nettles, fireweed, heaven knows what

else. Such a flagrant invitation to the abominable fifth column deeply shocked my father and his neighbours; and all their sympathy promptly shifted to this Quisling's immediate neighbours, now under constant paratroop invasion from the seeded composites and willow-herbs. I passed this derelict horror one cold winter day and to my joy saw one of Britain's rarest and most beautiful birds, a waxwing, happily feeding among a massive crop of berries on a tree there. But that was only a tiny poetic revenge.

Most of us remain firmly medieval, self-distancing and distanced from what we can neither own nor fully control, and from what we cannot see or understand. Just as the vast bulk of science fiction has decreed that anything that visits us from outer space must (in defiance of all probability) come with evil intent, so do we still assess most of nature, or at least where it comes close to us. Some deep refusal to accept the implications of Voltaire's famous sarcasm about the wickedness of animals in defending themselves when attacked still haunts the common unconscious; what is not clearly for mankind must be against it. We cannot swallow the sheer indifference, the ultrahumanity, of so much of nature. We may deplore the deforestation of the Amazon basin, the pollution of our seas and rivers, the extermination of the whale family and countless other crimes committed against the wild by contemporary man. But like nature itself, most of these things take place outside our direct knowledge and experience, and we seem incapable of supposing that responsibility for them (or lack of responsibility) might begin much closer to home, and in our own species' frightened past quite as much as in its helpless present—above all in our eternal association of ignorance with fear. I do not know how else one accounts for the popularity of such recent and loathsome manifestations of a purely medieval mentality as the film *Jaws*, and all its unhappy spawn.

The threat to us in the coming millennium lies not in nature seen as rogue shark, but in our growing emotional and intellectual detachment from it—and I do not think the remedy lies solely in the success or failure of the conservation movement. It lies as much in our admitting the debit side of the scientific revolution, and especially the changes it has effected in our modes of perceiving and of experiencing the world as individuals.

Science is centrally, almost metaphysically, obsessed by general truths, by classifications that stop at the species, by functional laws whose worth is valued by their universality; by statistics, where a Bach or a Leonardo is no more than a quotum, a hole in a computer tape. The scientist has even to generalize himself, to subtract all personal feeling from the conduct of experiment and observation and from the enunciation of its results. He may study individuals, but only to help establish more widely applicable laws and facts. Science has little time for minor exceptions. But all nature, like all humanity, is made of minor exceptions, of entities that in some way, however scientifically disregardable, do not conform to the general rule. A belief in this kind of exception is as central to art as a belief in the utility of generalization is to science; indeed one might almost call art that branch of science which present science is prevented, by its own constricting tenets and philosophies (that old *hortus conclusus* again), from reaching.

I see little hope of any recognition of this until we accept three things about nature. One is that knowing it fully is an art as well as a science. The second is that the heart of this art lies in our own personal nature and its relationship to other nature; never in nature as a collection of 'things' outside us. The last is that this kind of knowledge, or relationship, is not reproducible by any other means—by painting, by photography, by words, by science itself. They may encourage, foster and help induce the art of the relationship; but they cannot reproduce it, any more than a painting can reproduce a symphony, or the reverse. Ultimately they can only serve as an inferior substitute, especially if we use them, as some people use sexual relationships, merely to flatter and justify ourselves.

There is a deeper wickedness still in Voltaire's unregenerate animal. It won't be owned, or more precisely, it will not be disanimated, unsouled, by the manner in which we try to own it. When it is owned, it disappears. Perhaps nowhere is our human mania for possessing, our delusion that what is owned cannot have a soul of its own, more harmful to us. This disanimation justified all the horrors of the African slave trade. If the black man is so stupid that he can be enslaved, he cannot have the soul of a white man, he must be mere animal. We have yet to cross the threshold of emancipating mere animals; but we should not forget what began the emancipation of the slaves in Britain and America. It was not science or scientific reason, but religious conscience and fellow-feeling.

Unlike white sharks, trees do not even possess the ability to defend themselves when attacked; what arms they sometimes have, like thorns, are static; and their size and immobility means they cannot hide. They are the most defenceless of creation in regard to man, universally placed by him below the level of animate feeling, and so the most prone to destruction. Their main evolutionary defence, as with many social animals, birds and fishes, lies in their innumerability, that is, in their capacity to reproduce—in which, for trees, longevity plays a major part. Perhaps it is this passive, patient nature of their system of self-preservation that has allowed man, despite his ancient fears of what they may harbour in terms of other creatures and the supernatural, to forgive them in one aspect, to see something that is also protective, maternal, even womb-like in their silent depths.

All through history trees have provided sanctuary and refuge for both the justly and the unjustly persecuted and hunted. In the wood I know best there is a dell, among beeches, at the foot of a chalk cliff. Not a person a month goes there now, since it is well away from any path. But three centuries ago it was crowded every Sunday, for it is where the Independents came, from miles around along the border of Devon and Dorset, to hold their forbidden services. There are freedoms in woods that our ancestors perhaps realized more fully than we do. I used this wood, and even this one particular dell, in *The French Lieutenant's Woman*, for scenes that it seemed to me, in a story of self-liberation, could have no other setting.

This is the main reason I see trees, the wood, as the best analogue of prose fiction. All novels are also, in some way, exercises in attaining freedom—even when, at an extreme, they deny the possibility of its existence. Some such process of retreat from the normal world—however much the theme and surface is to be of the normal world—is inherent in any act of artistic creation, let alone that specific kind of writing that deals in imaginary situations and characters. And a part of that retreat must always be into a 'wild', or ordinarily repressed and socially hidden, self: into a place always a complexity beyond daily reality, never fully comprehensible or explicable, always more potential than realized; yet where no one will ever penetrate as far as we have. It is our passage, our mystery alone, however miserable the account that is brought out for the world to see or hear or read at second-hand.

The artist's experience here is only a special—unusually prolonged and self-conscious—case of the universal individual one. The return to the green chaos, the deep forest and refuge of the unconscious is a nightly phenomenon, and one that psychiatrists—and torturers—tell us is essential to the human mind. Without it, it disintegrates and goes mad. If I cherish trees beyond all personal (and perhaps rather peculiar) need and liking of them, it is because of this, their natural correspondence with the greener, more mysterious processes of mind—and because they also seem to me the best, most revealing messengers to us from all nature, the nearest its heart.

No religion is the only religion, no church the true church; and natural religion, rooted in love of nature, is no exception. But in all the long-cultivated and economically exploited lands of the world our woodlands are the last fragments of comparatively unadulterated nature, and so the most accessible outward correlatives and providers of the relationship, the feeling, the knowledge that we are in danger of losing: the last green churches and chapels outside the walled civilization and culture we have made with our tools. And this is however far we may have fled, or evolved away from knowledge of, attachment to, interest in the wild, or use of its imagery to describe our more hidden selves and mental quirks.

To see woods and forests merely scientifically, economically, topographically or aesthetically— not to understand that their greatest utility lies not in the facts derivable from them, or in their timber and fruit, or their landscape charm, or their utility as subject-matter for the artist— proves the gathering speed with which we are retreating into outer space from all other life on this planet.

Of course there are scientists who are aware of this profoundest and most dangerous of all our alienations, and warn us of it; or who see hopes in a rational remedy, in more education and knowledge, in committee and legislation. I wish them well in all of that, but I am a pessimist; what science and 'reason' caused, they cannot alone cure. As long as nature is seen as in some way outside us, frontiered and foreign, *separate,* it is lost both to us and in us. The two natures, private and public, human and non-human, cannot be divorced; any more than nature, or life itself, can ever be truly understood vicariously, solely through other people's eyes and knowledge. Neither art nor science, however great, however profound, can ultimately help.

I pray my pessimism is exaggerated, and we shall recover from this folly of resenting the fact that we are to all practical intents and purposes caged on our planet; of pretending that our life on it is a temporary inconvenience in a place we have outgrown, a boarding-house we shall soon be leaving, for whose other inhabitants and whose contents we need have neither respect nor concern. Scientists speak of biological processes recreated in the laboratory as being done *in vitro*; in glass, not in nature. The evolution of human mentality has put us all *in vitro* now, behind the glass wall of our own ingenuity.

There is a spiritual corollary to the way we are currently deforesting and denaturing our planet.

In the end what we must most defoliate and deprive is ourselves. We might as soon start collecting up the world's poetry, every line and every copy, to burn it in a final pyre; and think we should lead richer and happier lives thereafter.

W e park by a solitary row of granite buildings. To the east and behind it is a small half-hidden valley with two tall silent chimneys and a dozen or so ruined stone sheds, scattered about a long meadow through which a stream runs. The valley is bowered, strangely in this most desolate of Southern English land-scapes, by beech trees. Its ruins are now almost classical in their simplicity and seeming antiquity —and one is truly old, a medieval clapper bridge, huge slabs of rock spanning the little stream. But the rest were not designed, nor the beeches planted, to be picturesque. In Victorian times gunpowder for quarry-blasting was made and stored here. The stone sheds and chimneys were scattered, the trees introduced, the remote site itself picked, for purely safety reasons. Most contemporary visitors to Powder Mill Farm, on the southern fringe of the barren, treeless wastes of northern Dartmoor, are industrial archaeologists, summoned by this absurdly—in regard to its former use—Arcadian and bosky little valley behind. But we are here for something far more ancient and less usual still.

We set off north-west across an endless fen and up towards a distant line of tors, grotesque outcrops of weather-worn granite. Though it is mid-June, the tired grass is still not fully emerged from its winter sleep; and the sky is also tired, a high grey canopy, with no wind to shift or break it. What flowers there are, yellow stars of tormentil, blue and dove-grey sprays of milkwort, the delicate lilac of the marsh violet in the bogs, are tiny and sparse. Somewhere in the dark and uninhabited uplands to the north a raven snores. I search the sky, but it is too far off to be seen.

We cross a mile of this dour wasteland, then up a steep hillside, through a gap in an ancient sheep-wall, and still more slope to climb; and come finally to a rounded ridge that leads north to an elephantine tower, a vast turd of primary

rock, Longford Tor. At our feet another bleak
valley, then a succession, as far as the eye can
see, of even bleaker tor-studded skylines and
treeless moorland desert. My wife tells me I must
have the wrong place, and nothing in the landscape denies her. I do, but not with total convic-
tion. It is at least thirty years since I was last in
this part of the Moor.

We walk down the convex slope before us,
into the bleak valley, and I begin to think that
it must indeed be the wrong place. But then
suddenly, like a line of hitherto concealed infantry, huddled under the steepest downward fall
of the slope near the bottom, what we have come
for emerges from the low grass and ling: a thin,
broken streak of tree-tops, a pale arboreal surf.
For me this secret wood, perhaps the strangest
in all Britain, does not really rise like a line of
infantry. It rises like a ghost.

I can't now remember the exact circumstances
of the only other time I saw it, except that it
must have been late in 1946, when I was a
lieutenant of marines in a camp on the edge of
Dartmoor. This was not part of our training area,
and I can't have been on duty. It was winter,

there was ice in the air and a clinging mist, and I was alone. I think I had been walking somewhere else, trying to shoot snipe, and had merely made a last-minute detour to see the place, perhaps to orient myself.

At least it lived up to the reputation that I had once heard a moorland farmer give it: some tale of an escaped prisoner from Princetown a few miles away, found frozen to death there—or self-hanged, I forget. But it had no need of that kind of black embroidery. It was forlorn, skeletal, almost malevolent—distinctly eerie, even though I am not a superstitious person and solitude in nature has never frightened me one-tenth as much as solitude in cities and houses. It simply felt a bad place, not one to linger in, and I did not go into the trees; and I had never gone back to it, though often enough on Dartmoor, till this day. In truth I had forgotten about it, in all those intervening years, until I began writing this text and was recalling my father's suspicion of the wild. One day then its memory mysteriously surged, as it surges itself from the moorland slope, out of nowhere. Its name is Wistman's Wood.

I do not know who Wistman was—whether he was some ancient owner or whether the word derives from the old Devonshire dialect word *wisht*, which means melancholy and uncanny, wraithlike; and which lies behind one of Conan Doyle's most famous tales. There would never have been a hound of the Baskervilles, were it not for the much older Wisht Hounds of Dartmoor legend.

Wistman's Wood may be obscurely sited, but it is no longer, as it was in the 1940s, obscurely known. The rise of ecology has seen to that. In scientific terms it is an infinitely rare fragment of primeval forest, from some warmer phase of world climate, that has managed to cling on—though not without some remarkable adaptations—in this inhospitable place; and even more miraculously managed to survive the many centuries of human depredation of anything burnable on the Moor. Culturally it is comparable with a great Neolithic site: a sort of Avebury of the tree, an *Ur*-wood. Physically it is a half-mile chain of copses splashed, green drops in a tachist painting, along what on Dartmoor they call a clitter, a broken debris of granite boulders—though not at all on true tachist principle, by chance. These boulders provide the essential protection for seedlings against bitter winter winds and grazing sheep. But the real ecological miracle of Wistman's Wood is botanical. Its dominant species, an essentially lowland one, should not really be here at all, and is found at this altitude in only one other, and Irish, site in the British Isles. Here and there in the wood are a scatter of mountain ashes, a few hollies. But the reigning tree is the ancient king of all our trees: *Quercus robur*, the Common, or English, Oak.

We go down, to the uppermost brink. Names, science, history . . . not even the most adamantly down-to-earth botanist thinks of species and ecologies when he or she first stands at Wistman's Wood. It is too strange for that. The normal full-grown height of the common oak is thirty to forty metres. Here the very largest, and even though they are centuries old, rarely top five metres. They are just coming into leaf, long after their lowland kin, in every shade from yellow-green to bronze. Their dark branches grow to an extraordinary extent laterally; are endlessly angled, twisted, raked, interlocked, and reach quite as much downward as upwards. These trees are inconceivably different from the normal habit of their species, far more like specimens from a natural bonzai nursery. They seem, even though the day is windless, to be writhing, convulsed, each its own Laocoön, caught and frozen in some fanatically private struggle for existence.

The next thing one notices is even more extraordinary, in this Ice Age environment. It is a paradoxically tropical quality, for every lateral branch, fork, saddle of these aged dwarfs is densely clothed in other plants—not just the tough little polypodies of most deciduous woodlands, but large, elegantly pluming male ferns; whortleberry beds, grasses, huge cushions of moss and festoons of lichen. The clitter of granite boulders, bare on the windswept moors, here provides a tumbling and chaotic floor of moss-covered mounds and humps, which add both to the impression of frozen movement and to that of an astounding internal fertility, since they seem to stain the upward air with their vivid green. This floor like a tilted emerald sea, the contorted trunks, the interlacing branches with their luxuriant secondary aerial gardens . . . there is only one true epithet to convey the first sight of Wistman's Wood, even today. It is fairy-like. It corresponds uncannily with the kind of setting artists like Richard Dadd imagined for that world in Victorian times and have now indelibly given it: teeming, jewel-like, self-involved, rich in secrets just below the threshold of our adult human senses.

We enter. The place has an intense stillness, as if here the plant side of creation rules and even birds are banned; below, through the intricate green gladelets and branch-gardens, comes the rush of water in a moorland stream, one day to join the sea far to the south. This water-noise, like the snore of the raven again, the breeding-trill of a distant curlew, seems to come from another world, once one is inside the wood. There are birds, of course . . . an invisible hedgesparrow, its song not lost here, as it usually is, among all the sounds of other common garden birds, nor lost in its own ubiquity in Britain; but piercing and peremptory, individual, irretrievable; even though, a minute later, we hear its *prestissimo* bulbul shrill burst out again. My wood, my wood, it never shall be yours.

Parts of all the older trees are dead and decayed, crumbling into humus, which is why, together with the high annual humidity, they carry their huge sleeves of ferns and other plants. Some are like loose brassards and can be lifted free and replaced. The only colour not green or bronze or russet, not grey trunk or rich brown of the decaying wood, are tiny rose-pink stem-beads, future apples where some gall-wasp has laid its eggs on a new shoot. But it is the silence, the waitingness of the place, that is so haunting; a quality all woods will have on occasion, but which is overwhelming here—a drama, but of a time-span humanity cannot conceive. A pastness, a presentness, a skill with tenses the writer in me knows he will never know; partly out of his own inadequacies, partly because there are tenses human language has yet to invent.

We drift from copse to copse. One to the south is now fenced off by the Nature Conservancy to see what effect keeping moorland sheep, bullocks and wild ponies from grazing will have. It has a much denser growth at ground level, far more thickety, and is perhaps what the wood would have looked like centuries ago, before stock was widely run on the Moor; and yet now seems artificial—scientifically necessary, aesthetically less pleasing, less surreal, historically less honest beside the still open wood, 'gardened' by what man has introduced. There is talk now of wiring off the whole wood like this, reserving it from the public, as at Stonehenge. Returning, we come on two hikers, rucksacks beside them, lying on their backs inside the trees, like two young men in a trance. They do not speak to us, nor we to them. It is the place, wanting it to oneself, and I am prey to their same feeling. I persuade my wife to start the long climb back. I will catch up.

I go alone to the most detached and isolated of the copses, the last and highest, to the north. It grows in a small natural amphitheatre, and proves to be the most luxuriant, intricate and greenly beautiful of the chain. I sit in its silence, beneath one of its most contorted trees, a patriarchal gnome-oak. The botanist in me notices a colony of woodrush, like a dark green wheat among the emerald clitter; then the delicate climbing fumitory *Corydalis claviculata*, with its maidenhair-fern leaves and greenish-white flowers. A not uncommon plant where I live in Dorset; yet now it seems like the hedgesparrow's song, hyperdistinct, and also an epitome, a quintessence of all my past findings and knowledge of it; as with the oaks it grows beneath, subsuming all other oaks. I remember another corydalis, *bulbosa*, that they still grow in the garden at

Uppsala in honour of the great man, who named the genus.

From somewhere outside, far above, on top of Longford Tor, I hear human voices. Then silence again. The wood waits, as if its most precious sap were stillness. I ask why I, of a species so incapable of stillness, am here.

I think of a recent afternoon spent in discussion with a famous photographer, and how eminently French and lucid his philosophy of art seemed, compared to mine. I envied him a little, from the maze of my own constantly shifting and confused feelings. I may pretend in public that they are theories, but in reality they are as dense and ravelled as this wood, always beyond my articulation or rational comprehension, perhaps because I know I came to writing through nature, or exile from it, far more than by innate gift. I think of my father and, wrily, of why I should for so many years have carried such a bad, unconsciously repressing mental image of Wistman's Wood—some part or branch of him I had never managed to prune out. It is incomprehensible now, before such inturned peace, such profound harmlessness, otherness, selflessness, such unusing . . . all words miss, I know I cannot decribe it.

A poet once went near, though in another context: *the strange phosphorus of life, nameless under an old misappellation.*

So I sit in the namelessness, the green phosphorus of the tree, surrounded by impenetrable misappellations. I came here really only to be sure; not to describe it, since I cannot, or only by the misappellations; to be sure that what I have written is not all lucubration, study dream, *in vitro*, as epiphytic upon reality as the ferns on the branches above my head.

It, this namelessness, is beyond our science and our arts because its secret is being, not saying. Its greatest value to us is that it cannot be reproduced, that this being can be apprehended only by other present being, only by the living senses and consciousness. All experience of it through surrogate and replica, through selected image, gardened word, through other eyes and minds, betrays or banishes its reality. But this is nature's consolation, its message, and well beyond the Wistman's Wood of its own strict world. It can be known and entered only by each, and in its now; not by you through me, by any you through any me; only by you through yourself, or me through myself. We still have this to learn:

the inalienable otherness of each, human and non-human, which may seem the prison of each, but is at heart, in the deepest of those countless million metaphorical trees for which we cannot see the wood, both the justification and the redemption.

I turned to look back, near the top of the slope. Already Wistman's Wood was gone, sunk beneath the ground again; already no more than another memory trace, already becoming an artefact, a thing to use. An end to this, dead retting of its living leaves.

About the Author

John Fowles is widely regarded as one of the greatest English novelists of the twentieth century. Born in Leigh-on-Sea in Essex, England, he won international recognition with his first novel, *The Collector*, in 1963. His many other bestselling novels include *The Magus* (1966), *Daniel Martin* (1977), and *The French Lieutenant's Woman* (1969), which was turned into an acclaimed film starring Meryl Streep and Jeremy Irons. John Fowles died in 2005.